IMAGES
of Rail

CAMAS PRAIRIE RAILROAD

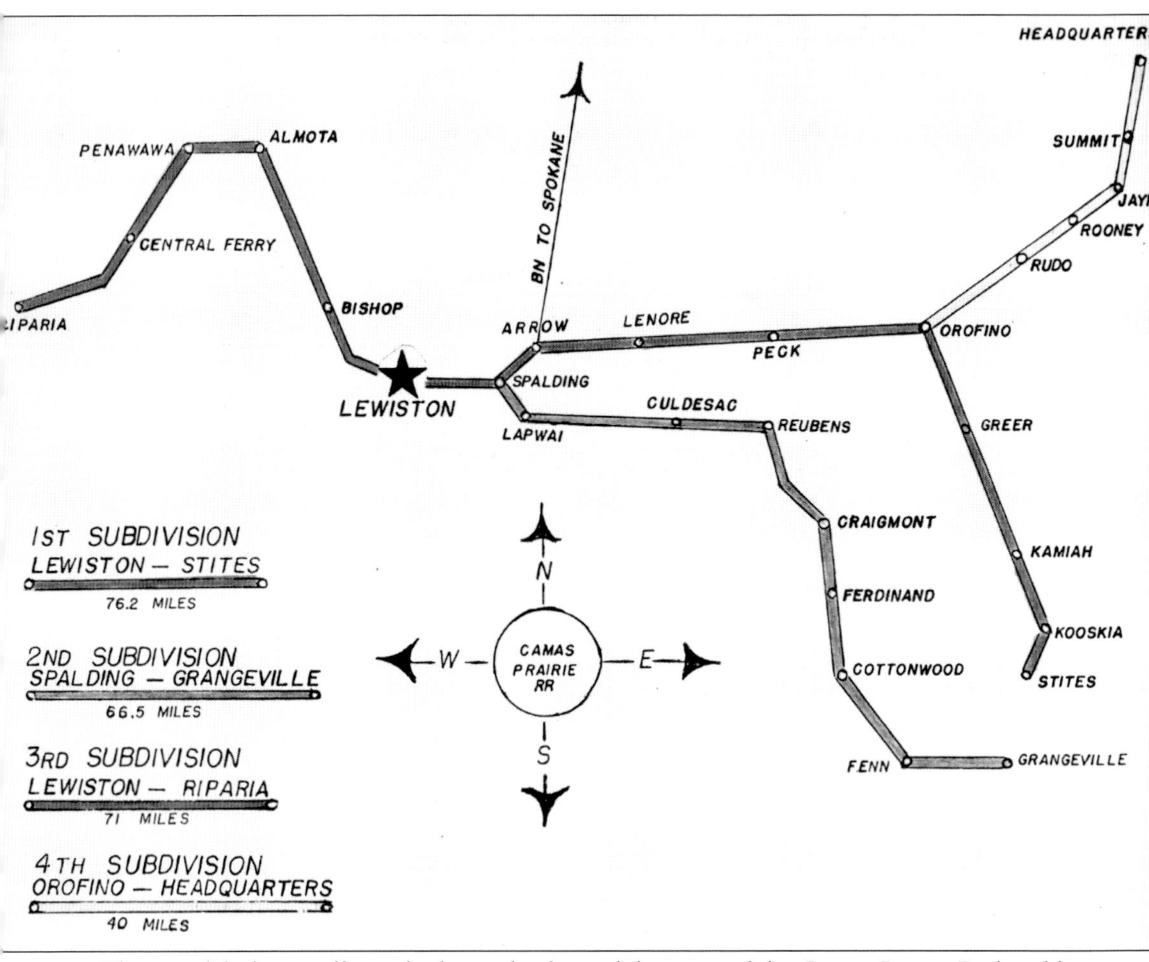

This simplified map effectively shows the four subdivisions of the Camas Prairie Railroad but does little to express the roller-coaster-like twists and turns and unique terrain faced by each branch. The first branch followed along the Clearwater River, the second through steep canyons onto a prairie, the third along the Snake River through the arid Channeled Scablands, and the fourth through the forest. (Courtesy of the University of Idaho Special Collection and Archives, MG183-F06-016-001.)

On the Cover: One of the first major hurdles of the second subdivision of the Camas Prairie Railroad, Lapwai Canyon proved difficult and required no less than 7 tunnels and 19 bridges in an eight-mile stretch. Timber for the bridges was cut at sawmills conveniently located along the line. (Courtesy of the University of Idaho Special Collection and Archives, MG183-F06-016-001.)

Robert Perret and Amy Thompson

ISBN 978-1-4671-0770-9

Published by Arcadia Publishing
Charleston, South Carolina

Printed in the United States of America

Library of Congress Control Number: 2021946733

For all general information, please contact Arcadia Publishing:
Telephone 843-853-2070
Fax 843-853-0044
E-mail sales@arcadiapublishing.com
For customer service and orders:
Toll-Free 1-888-313-2665

Visit us on the Internet at www.arcadiapublishing.com

Contents

ACKNOWLEDGMENTS

We would like to thank our editor Caroline (Anderson) Vickerson for her support and guidance as we created this book. The images in this volume were collected from the University of Idaho Special Collections and Archives; in particular, they are from the A.B. Curtis Collection (PG 13), Ron V. Nixon Collection (PG 16), Hal Riegger Papers (MG 183), Idaho Cities and Towns Photographs (PG 5), Idaho Photographs (PG 6), Clifford M. Ott Collection (PG 90), Kyle Laughlin Photographs (PG 99), Joel King Photographs (MA 2016-18), Craigmont Vollmer Photographs (MA 2021-29), Potlatch Corporation Historical Archives (MG 457), Stephen Shawley Collection (PG 38), Ott-Hodgins Photograph Collection (PG 91), and Northwest Historical Postcard Collection (PG 9).

Introduction

The story of the Camas Prairie Railroad begins at the start of the 20th century. It was an age of railroad barons, who were racing against each other to control territory by being the first to lay down tracks. The Union Pacific (UP) and Northern Pacific (NP) were vying to control the Pacific Northwest. UP seemed to have a head start, with a line running through Boise, Idaho, and on to Portland, Oregon. In 1898, UP had taken control of the Oregon Railroad and Navigation Company, giving the company a vested interest in connecting Oregon to the rest of the country. NP had a line running from Helena, Montana, to Tacoma, Washington, by way of Sandpoint, Idaho. Between them lay a tantalizing blank space on the rail line maps—the heart of the Idaho panhandle and the most direct route through much of the inland northwest.

NP had begun the painstaking process of crossing the Camas Prairie from the east, connecting Spalding to Grangeville, Idaho, with a circuitous 66-mile short line. UP had connected Riparia, Washington, with Lewiston, Idaho, and the Snake River. Roughly 75 miles separated the rivals, with UP officials longingly looking at the commerce of the east and NP officials seeing the resources of the Pacific Northwest coast just over the horizon. For each railroad, that last stretch of rolling hills and rugged valleys seemed insurmountable. And so, in the wilds of Idaho, an unprecedented deal was struck. The two railroads agreed to form a third company—which would be shared equally—to bridge the Camas Prairie. Each company halved its risk and had much to gain.

On December 3, 1909, the Camas Prairie Railroad (CPR), headquartered in Lewiston, went into operation. At that time, the CPR owned no trains—only the soaring, twisting tracks that towered above and passed through the rugged landscape. In some places, the trestles loom 50 feet above the small towns they pass over, which earned the CPR a nickname: "the railroad on stilts." During its 92-year history, the CPR primarily served as a freight line for the products of the Northwest, including lumber, ore, and agricultural products, allowing the Inland Empire to flourish.

Where they had once been trying to each build a railway around the other, NP and UP now began to connect their lines from NP's new line through the Lapwai Canyon near Culdesac, Idaho, down to UP's route between Riparia and Lewiston. Although the Nezperce & Idaho Railroad was technically an independent railway, we are following the established practice of treating it as part of the CPR, since it is effectively a stub that connects the city of Nezperce to the railway. Rails were run to the major lumber towns in the region, and wood soon made up over 70 percent of the freight carried upon the rails. The rest was largely other agricultural products, despite early efforts to bolster passenger service. In the last decades of the CPR, wheat had displaced lumber as the predominant freight. Ultimately, the CPR spanned 267.7 miles, connecting the lumber camps on the inland side of Idaho to the Snake River in Lewiston, and from there to the Pacific Ocean and the rest of the world.

The Great Depression caused the demand for lumber to plummet, leading many of the sawmills along the railroad—and the towns they supported—to falter; many of the lumber companies consolidated into the Potlatch Lumber Company to survive. Others paused operations for a few years or completely went out of business. At the same time, demand for passenger service almost vanished, and the railroad invested in relatively inexpensive electric locomotives to try to salvage what it could of the commuter business. Passenger service ended for many of the rural, outlying stations as early as 1955, unable to compete with the highways that had begun to creep across the landscape. Business rebounded during World War II with a sudden demand for building materials. In the decade after that war, a housing boom continued to fuel a demand for lumber.

The remains of the railroad, much of which can be accessed along US Route 95, are still breathtaking. In particular, the 17-mile run through Lapwai Canyon and across the Halfmoon Bridge remains a favorite destination for sightseeing. In just 13 miles, the track passes through 7 tunnels and 17 wooden trestles, culminating in the famous curving bridge that towers 141 feet above the forested hills. Engineers reported that the bridge would sway as much as eight feet as trains passed. Likewise, the Horseshoe Tunnel was so long and close that exhaust from the engines would make the air almost unbreathable. Just outside of Grangeville was an incline that often required two engines and grit on the tracks to provide extra traction in order for trains to surmount the hill.

The end of the Camas Prairie Railroad began in 1985, when the tracks between the small towns of Revling and Headquarters and Kooskia and Stites were abandoned. The original railways, including towering bridges and hairpin tunnels, had been constructed from the same lumber they carried, and the wood had simply reached the end of its useful life. To replace the old track with modern steel was prohibitively expensive. In 1997, the railway was sold to North American Railnet, a holding company that invested in short line railroads. In 2000, the easternmost segment, from Spalding to Grangeville, was also taken out of service. In 2004, the railway was sold to Watco Companies LLC, which specializes in operating short line railroads. However, Watco only actively used the original line between Riparia and Lewiston, leasing access to the rest of the tracks that remained serviceable. Weather and fire have continued to take a toll on the wooden lines, and tracks continue to be removed when they are found to be unsafe.

The story of the Camas Prairie Railroad is one of railways in the Northwest, as it passed from one owner to another and moved through time from steam-powered locomotives to electric trains. The railway united communities such as Spokane, Lewiston, Lapwai, Spalding, Craigmont, and Grangeville. As readers will discover while viewing the photographs within this volume, the Camas Prairie is beautiful and dynamic, and the railroad's remarkable tracks, bridges, and tunnels are truly inspiring.

One

A Truce in the Great Railroad War

When the Native American reservations of the inland northwest were opened to white settlement in 1895, a rush of developers began competing to build and control rail routes through the region. The rolling hills and steep valleys of the Camas Prairie proved to be a special challenge, however, prompting two competitors—Edward Harriman, president of Union Pacific, and James Hill, chief executive officer of the Great Northern Railway—to cooperate in building a new route.

From the start, efforts to reach Lewiston by rail must have seemed cursed, because weather catastrophes, economic downturns, steep terrain, and the inability to secure rights-of-way halted production for almost two decades. The UP approached from the west from Portland along the Columbia and Snake Rivers, making it all the way to Riparia (about 75 miles from Lewiston) before construction was halted due to the Panic of 1893. Steamboats continued to carry commodities from Lewiston to Riparia until the rail line could be completed, and this effective yet limited mode of transportation did not lend to the urgency to complete the railroad. Around the same time, coming from the east, the NP trudged through northern Idaho before turning south to capitalize on the promised prosperity offered by Lewiston and the Camas Prairie. The NP made it as far as Genesee, Idaho (11 miles away), before the steep 2,000-foot drop in elevation forced it to find another way.

The competition between the two railroads was the catalyst that reignited railroad construction. Both companies quickly realized that for them to get to Lewiston and continue to Grangeville, thus taking advantage of the agricultural potential, they would have to run lines directly next to each other in the narrow and steep valley. So, they called a truce, and the Camas Prairie Railroad was born.

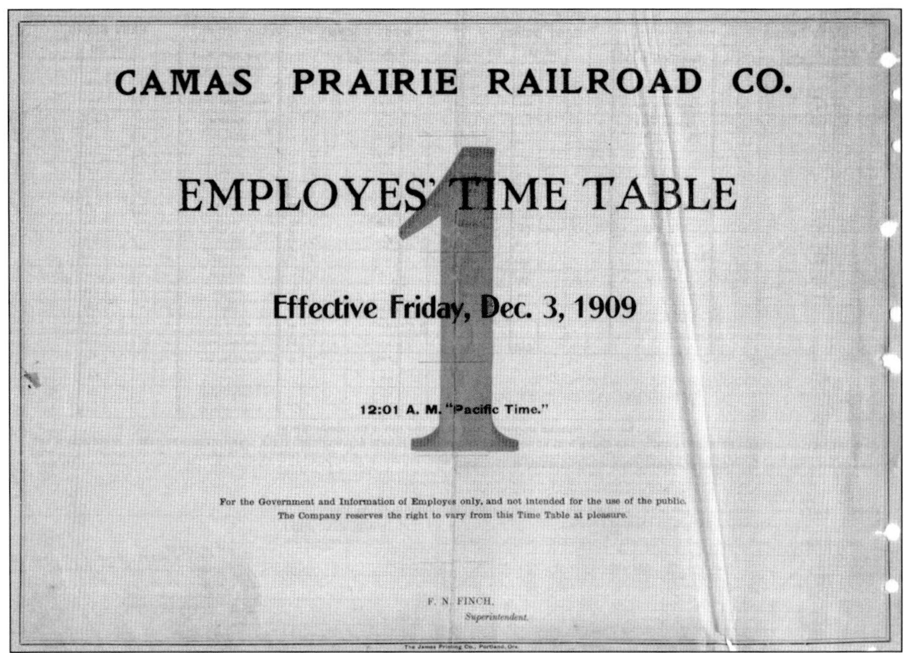

CAMAS PRAIRIE RAILROAD CO.

EMPLOYES' TIME TABLE

1

Effective Friday, Dec. 3, 1909

12:01 A. M. "Pacific Time."

For the Government and Information of Employes only, and not intended for the use of the public.
The Company reserves the right to vary from this Time Table at pleasure.

F. N. FINCH,
Superintendent.

The James Printing Co., Portland, Ore.

Updated every year, the employee timetables gave instructions and rules regarding the movement of all trains and equipment running on Camas Prairie Railroad lines. Other essential information included station names and the exact times trains were expected to arrive, as well as the amenities found at each destination. The rules included the limits of freight each train could carry. Later timetables also listed rules governing employees. One interesting thing to note is that the station names have changed—for example, Lewiston Junction was the name of the station in a town formerly known as Texas City, but in 1911, both the station and town adopted the name Riparia. (Above, MA 2016-25-137-001; below, MA 2016-25-137-002.)

FIRST DISTRICT — Time Table No. 1, Effective Dec. 3, 1909.

West Bound: Third Class 857 Freight Ex. Sun.	West Bound: First Class 243 Passenger Daily	Water, Coal, Scales, Tables and Wys.	Station Numbers	Distance from Grangeville	Stations (Telegraph Offices and Calls)	Distance from Joseph	Capacity of Side Tracks	East Bound: First Class 244 Passenger Daily	East Bound: Third Class 858 Freight Ex. Sun.
AM Lv 7.10	AM Lv 7.00	W C Y	C P 146	0.0	Gv GRANGEVILLE D	66.4	60	PM Ar 6.45	PM Ar 4.30
					6.9				
7.45	7.20		C P 142	6.9	FENN	59.5	35	6.20	3.45
					8.7				
8.45	7.50		C P 133	15.6	Co COTTONWOOD	50.8	50	5.55	3.00
					8.1				
9.45	8.20	W 3 Mi E	C P 126	23.7	STEUNENBERG	42.7	20	5.30	2.15
					8.2				
10.45	8.50		C P 117	31.9	Vo VOLLMER D	34.5	30	5.05	1.30
					8.3				
PM 12.01	9.15	Y W	C P 109	40.2	Ru REUBENS D	26.2	30	4.40	12.45 / 12.01
					14.5				
1.15	10.10	W 8 Mi E	C P 94	54.7	CU CULDESAC D	11.7	50	3.30	10.10 / 9.50
					6.4				
1.45	10.35		C P 88	61.1	SWEETWATER	5.3	20	3.05	9.15
					2.2				
1.55	10.40		C P 86	63.3	FORT LAPWAI	3.1	12	2.48	8.45
					3.1				
2.10 PM Ar	10.55 AM Ar	Y	C P 80	66.4	JOSEPH	0.0	No Sdg.	PM Lv 2.40	AM Lv 8.35
Ex. Sun.	Daily							Daily	Ex. Sun.
7.00	3.55				Time over District			4.05	8.05
8.9	16.8				Average Speed per Hour			16.3	8.2
C. & P. Ry. Time Table	C. & P. Ry. Time Table							C. & P. Ry. Time Table	C. & P. Ry. Time Table

SECOND DISTRICT — Time Table No. 1, Effective Dec. 3, 1909.

West Bound: First Class 5 Passenger Daily	Water, Coal, Scales, Tables and Wyes	Station Numbers	Distance from Lewiston	Stations (Telegraph Offices and Calls)	Distance from Lewiston Junction	Capacity of Side Tracks	East Bound: First Class 6 Passenger Daily
PM Lv 6.00	W C T	C P 73	0.0	Wn LEWISTON	72.0	125	AM Ar 7.40
				1.0			
6.03		C P 71	1.0	TRANSFER	71.0	15	7.37
				5.1			
6.12		C P 66	6.1	DAISY	65.9	70	7.24
				3.3			
6.21		C P 62	9.4	ALPOWA	62.6		7.17
				0.9			
6.23		C P 62	10.3	MOSES	61.7	71	7.15
				6.8			
6.39	W	C P 55	17.1	INDIAN	54.9	72	7.01
				5.4			
6.50		C P 50	22.5	BISHOP	49.5	71	6.49
				1.8			
6.54		C P 48	24.3	TRUAX	47.7		6.45
				3.8			
7.03		C P 44	28.1	WAWAWAI	43.9	77	6.36
				2.0			
7.07		C P 42	30.1	INTERIOR	41.9	19	6.32
				5.9			
7.20	W	C P 36	36.0	Al ALMOTA	36.0	71	6.19
				6.8			
7.36		C P 29	42.8	SWIFT	29.2	70	6.05
				5.3			
7.47		C P 24	48.1	PENAWAWA	23.9	70	5.53
				9.4			
8.08	W	C P 15	57.5	CENTRAL FERRY	14.5	71	5.32
				14.5			
PM Ar 8.45	C Y	C P 0	72.0	Ax LEWISTON JCT.	0.0	71	AM Lv 5.00
Daily				72.0			Daily
2.45				Time over District			2.40
26.2				Average Speed per Hour			27.0

East-bound Trains are superior to Trains of the same class in the opposite direction.

IMPORTANT.—ALL TRAINS BETWEEN LEWISTON AND JOSEPH WILL BE GOVERNED BY NORTHERN PACIFIC RAILWAY TIME TABLE, IDAHO DIVISION NUMBER 31 OF OCTOBER 31, 1909.

In the operation of the Camas Prairie Railroad employees will be governed by the Operating Department Rules and Regulations of the Northern Pacific Railway. Employees must provide themselves with a copy of the Book of Rules and Regulations of the Operating Department of the Northern Pacific Railway Co. and time table No. 31 of the Idaho Division of the Northern Pacific Railway Co. They will also provide themselves with copy of rules and current time table of the Washington Division of the Oregon Railroad & Navigation Co. and be governed by same in the use of terminals at Lewiston Jct.

Trains 5 and 6 will stop on flag to pick up or let off passengers about one mile east of Central Ferry where ferry crosses the river.

When sand is blowing engineers will run with great care and under control where they cannot see track is clear.

Mountain grade extends between Reubens and Sweetwater.

IMPORTANT: Special attention is called to rules 500 to 518, inclusive, regarding operation on mountain grades between Reubens and Sweetwater.

TONNAGE RATING OF FREIGHT ENGINES.

FIRST DISTRICT	Class E 1-2-3-4, F-1 A	Class E 1-2-3-4, F-1 B	Class F, F 4 A	Class F, F 4 B	Class E 1 A	Class E 1 B	Class E 2-3 D 1-3 A	Class E 2-3 D 1-3 B	Class B, B 1 A	Class B, B 1 B	Class C A	Class C B
East Bound.												
Joseph to Sweetwater	700	640	600	540	540	486	480	432	450	405	360	324
Sweetwater to Culdesac	500	450	400	350	350	300	300	250	250	200	200	150
Culdesac to Reubens	250	200	200	150	150	130	125	100	100	75	75	50
Reubens to Vollmer	950	900	800	750	700	650	650	600	575	525	525	475
West Bound.												
Vollmer to Reubens	950	900	800	750	750	700	700	650	625	575	550	500
Reubens to Culdesac						Twenty	Cars.					
Culdesac to Sweetwater						Thirty	Cars.					
Sweetwater to Joseph						Thirty	Cars.					

COMMERCIAL SPURS.

DISTANCE FROM JOSEPH.		Car Capacity
Caldwell's	8.0 Miles	6
Jacques	9.6 "	5
Gwyns	27.5 "	6
Clicks	29.6 "	8
Craig Mountain Ry.	30. "	4

List of Surgeons:
DR. J. B. MORRIS, Surgeon, Lewiston, Idaho.
DR. F. A. CAMPBELL, " Grangeville, "

Registering Stations:
Lewiston, Lewiston Jct.
Joseph and Grangeville.

Railroads offered the primary means to access lumber to the northeast and south of Lewiston. Starting in 1910, firms such as the Craig Mountain Lumber Company transferred lumber through the Camas Prairie Railroad from Winchester, Idaho. As the largest mill in northern Idaho, Craig Mountain Lumber Company employed 270 men, ensuring that the rail line ran twice a day carrying passengers and freight. (PG 6-55-73.)

The rolling hills of the Camas prairie presented a real challenge for westward expansion. Even the famed Lewis and Clark expedition nearly starved to death in the area. However, the Nez Perce tribe knew how to make flour from the plants and thrive on the Palouse. (PG 9-17-16a.)

Construction started on the Lawyers Canyon viaduct in 1906, with crews having to haul cement from the valley almost 3,000 feet up by wagon. The techniques used to get the concrete footings in place ranged from laborers carrying them down the steep sides of the canyon by hand to using a one-yard bucket dangling from a one-and-a-half-inch cable. The steel was carried and set in place by a derrick car that inched its way along the newly completed sections. (MG 183-F16-056.)

A similar derrick car as in the previous image inches its way across the deck of the Lawyers Canyon viaduct (shown here near completion) about halfway between Craigmont and Ferdinand, Idaho. It spans 1,500 feet and stands 287 tall, making it the tallest railroad bridge in the world. While no longer used, the viaduct still stands. (MG 183-F16-024.)

Track crews were required to complete miraculous and dangerous feats of labor at a time when few to no safety precautions existed. It was expected that between 2,000 and 3,000 men would work on the construction of the line, and they were stationed at temporary camps along the route. The best of the skilled laborers could make $8 per day in 1908. In this image, two men are lowering a wooden beam over the edge of bridge No. 39 near Pardee on the line heading to Stites, Idaho. Pardee, which is now a ghost town along the South Fork of the Clearwater River, was rumored to be home to a mistress and illegitimate son of Pres. Grover Cleveland. (MG 183-F16-042.)

Cammassia quamash, the common camas, is a perennial herb grown from a bulb with blue-violet flowers that once covered the upland prairies above the Snake River, which led to the region—and, incidentally, the railroad—being dubbed the camas prairie. In addition to being an ornamental plant that could fill whole fields with lovely blue flowers, the bulbs, once properly prepared, are edible and said to taste like baked pears, cooked figs, or a sweeter version of sweet potatoes. Being unused to the starchy texture, Lewis and Clark became quite ill after they were given the prepared bulbs to eat by the Nimiipuu (Nez Perce) people when they had nothing else to eat. (PG 99-867-1.)

The camas prairie region, which originally contained about 640 square miles of open grassland where the native camas plant grew, was turned into fertile farmland to support grains such as wheat and barley. The Camas Prairie Railroad was instrumental in transporting harvests from isolated barns and silos to market and also in bringing back supplies and equipment. (PG 99-46-023.)

The Camas Prairie Railroad was driven through the rolling hills of the camas prairie with little more than sheer determination. Many times, horse-drawn plows were the only option to pack and level the earth. Creating a smooth and solid underlayment for the track required a tremendous effort with no guarantee of success. In more welcoming places, eight miles of track could be laid in a day, but on the camas prairie, it took several years to cross a little more than 100 miles. (PG 90 1-6-73-007.)

Cable-driven tramways were used extensively throughout northern Idaho and eastern Washington to carry grain down the steep canyon walls. One stop on the Camas Prairie Railroad was labeled Tramway, although it never developed into a town; following the development of paved roads, it did not remain in existence for much longer. (PG 90-04-082.)

Building a railroad was not always a straightforward task. Sometimes, temporary structures and tracks had to be built in the remotest of areas in order for trains to travel far enough to carry supplies to where they were needed. Here, a temporary—and rather rickety-looking—trestle was built until the ground could be filled in to create a more stable foundation. (PG 13-12920.)

Even before the creation of the Camas Prairie Railroad, the Northern Pacific saw the value in building a line to the camas prairie. The location given for this photograph is Stuart, Idaho, which was renamed Kooskia just after 1900. W.T. Fields is the engineer in white sleeves, but the names of the other two men have been lost to time. (MG 183-F09-008.)

In remote areas, the construction materials for new tracks were often carried by trains traveling along the yet-to-be-completed line. Workers would build a section of track, then move the train forward, and this would be repeated until they reached the terminal point—or until roads, rivers, or other options presented themselves. (MG 183-F21-003.)

The camas prairie land had first seemed impassable to locomotives, but with the aid of dozens of wooden tresses like this one, the Union Pacific and Northern Pacific managed to lay level tracks from the heart of timber country to the Snake River, where lumber could be moved down to Oregon and California and then shipped across the United States. (MA 2016-18-001.)

A striking close-up of the Camas Prairie Railroad trestle near Ferdinand shows the intricate engineered design. Wooden trestle bridges came into fashion in the 1830s. Untreated lumber lasts only about 20 years, and the friction caused by the locomotives running over the track could set the wood on fire. However, these bridges were relatively quick to construct and much less expensive than steel trestles. (MG 183-F15-004.)

The Camas Prairie Railroad brought hope and prosperity to many in the small communities of rural north-central Idaho. People traveled from far and wide to get a glimpse of—or even ride on—the first train on the Grangeville line. Celebrations were held in many of the towns along the route, and around 100 people from Grangeville crowded into the train, which also picked up numerous people in other towns along the way to Lewiston. (MG 183-F26-004.)

The swells and dips of the camas prairie were formed during the last ice age, when the expansion and retreat of glaciers ground the underlying basalt and sandstone into fine granules called loess. Strong winds blowing over the wide expanses over thousands of years formed the dunes of today. Loess is one of the most productive agricultural soils in the world. (PG 99-366-02.)

The Camas Prairie Railroad did not always go over the rough terrain. The short line features many tunnels, which are infamous for their close dimensions and tight turns. In the 23-mile section between Culdesac and Craigmont, there are seven of these remarkable tunnels. (MG 183-F14-010.)

The westbound Grangeville local is shown way up on the Lawyers Canyon bridge. At 276 feet high and over 1,500 feet across, this is the widest and deepest canyon the railway faced. Construction on the Lawyers Canyon viaduct, designed by the American Bridge Company, lasted from 1906 to 1908; this was one of the last hurdles involved in completing the second subdivision of the Camas Prairie Railroad linking Lewiston to Grangeville. (MG 183-F10-010.)

Northern Pacific locomotive No. 1 waits at Reubens, Idaho, in 1912, appearing ready for its portrait to be taken. The prominent boilers and wood chips leave no doubt that this is a classic steam engine. This was one of the earliest locomotives to run on the Camas Prairie Railroad. (MG 183-F25-007.)

Workers load a railcar parked on the tracks at the western edge of the not-yet-completed Lawyers Canyon viaduct. The drop beneath them is nearly 300 feet. Neither they nor their furry friend seem to be fazed. The railcar would have been used to ferry construction tools and supplies to the men as they worked. (MG 183-F16-004.)

Even in the late 1920s, trestle construction was an arduous task when workers were building the fourth subdivision between Orofino and Headquarters. Hand tools were still in extensive use—these included the two-man cart used to move supplies and debris and shovels used to spread fill material to help support the bank. Despite the branch being only 41 miles long, it included 59 bridges, primarily through the Orofino Creek canyon. The total cost of construction was estimated at $3.6 million. (Above, PG 13-6126; below, PG 13-6127.)

This 1921 photograph shows a man on trestle timber bridge No. 46 near Cottonwood, Idaho. Cottonwood was founded in 1863 as the natural resting place for a wagon team traveling between Lewiston and Grangeville. For its first 30 years, it contained little more than a post office, blacksmith, carpenter, and saloon, making it the 19th-century version of a truck stop. (MG 183-F16-003.)

A fully loaded freight train makes its way through Lapwai Canyon between Lewiston and Grangeville. The newly paved US Route 95 is visible in the distance; it replaced the notorious Old Spiral highway, a steep switchback that accommodated the sudden 2,000-foot elevation change of the Winchester Grade in a road that was only 7.3 miles long. (PG 9-17-10a.)

Early wooden railroad bridges, like this one from the 1910s, had to be built by hand. Men balanced on beams used ropes and pulleys to hoist large timbers into place, while others looked on and prepared the next load. Most of the wooden materials were locally sourced and prepared in neighboring mills. (PG 6-62-28a.)

This postcard features the famous Half Moon Bridge, which is still regarded as one of the most scenic points on the Camas Prairie Railroad. It stands a remarkable 141 feet high despite being made of wooden beams. One million board feet of timber went into this impressive trestle. (MG 183-F14-009.)

Some dapper gentlemen are included among the train crew standing next to the No. 72 steam engine in the Lewiston railroad yards in the late 1890s or early 1900s. In anticipation of—or perhaps to entice the arrival of—railroad companies in Lewiston, a depot was completed in 1895; however, the first train did not arrive until 1898. (PG 91-629.)

Pictured in 1925 is a familiar sight on the Nez Perce leg, the No. 3, a local powerhouse for the first half of the 20th century, working as a common carrier until 1950. A common carrier locomotive is one that is for hire rather than being owned by a particular railroad. In fact, common carriers were prohibited from refusing to transport any legal freight or law-abiding person. (MG 183-F11-001.)

It is an amazing sight to see a locomotive pass 100 feet overhead. Steam trains in the early 20th century had an average traveling speed of approximately 50 miles per hour. On straight track laid on a solid bed, these trains could reach speeds of more than 120 miles per hour, but on an elevated track like this, they would have been more likely to slow down rather than speed up. (MG 183-F16-012.)

Steam locomotives, like this Shay locomotive No. 102, serviced the Potlatch Lumber Company. Spur lines branching into Potlatch's great lumber holdings connected to the Camas Prairie Railroad's fourth subdivision, bringing lumber to a worldwide market. (PG 6-62-1.)

The caboose of train No. 858 is parked opposite the water pump house and well derrick at Reubens. Pump houses were vital in the operation of steam trains, which actually required more water than coal. As the railway transitioned to coal-fired engines, the need for pump houses disappeared. (MG 183-F28-005.)

Northern Pacific No. 1521 is shown switching tracks to head toward Craigmont. Sitting on the Nez Perce reservation, the town of Craigmont was originally named Chicago in 1898. This apparently caused confusion for the postal service, so the name was changed to Ilo, and the town later merged with nearby Vollmer to become Craigmont. (MG 183-F25-022.)

In the remote wilderness of the camas prairie, railroads had to clear the snow from their own tracks. To do this, they would use rotary plows—essentially giant snowblowers attached to the front of a locomotive and powered by the drive of the engine itself. (MG 183-F09-012.)

This image of people standing at the mouth of a tunnel emphasizes the scope of the work that had to be done to make the camas prairie traversable by train. Tunnels had to be as much as 20 feet tall and 14 feet wide to allow trains to pass through them. (MG 183-F14-022.)

A barge prepares to move the central section of a swing bridge, which will allow a ship to pass. The swing bridge was the predecessor to the modern drawbridge, which allowed trains and boats to pass through the same shipping areas. (PG 90-10-014f.)

The old US Route 95 (or "north-south highway") ran along the edge of Lawyers Canyon, which was named after Hallalhotsoot, a Nimiipuu (Nez Perce) leader who received the nickname "Lawyer" due to his exceptional oratory skills, which led many to assume he was chief of the tribe. Lawyer's father, Twisted Hair, befriended the Lewis and Clark expedition and helped the explorers on their journey through the valley. Lawyers Creek winds through the center of the steep canyon. Pictographs found within the canyon show that people have inhabited the region for 10,000 years. The highway was completed in 1926, but due to the steep canyons and sharp curves, it was a dangerous route that had to be taken at a slow pace. (PG 99-686-003.)

This postcard shows one of the famed wooden trestles found at mile 31 along the second subdivision of the Camas Prairie Railroad on its way to Grangeville. The wooden bridge is 586 feet long and 132 feet high. Besides transporting lumber and grain, passenger cars brought people to and from otherwise inaccessible small towns along the way. (PG 6-47-1.)

Parked side by side are trains from the two competitors that came together to build the Camas Prairie Railroad—the Union Pacific and Northern Pacific. The owners of each railroad had tried to conquer the Camas Prairie on their own before they united to build a joint line that began service in 1908. They operated the railroad together for nearly 90 years before selling it in 1998. (MG 183-F13-004.)

This postcard shows the famous bridge or viaduct over Lawyers Canyon. At the time of completion, it was the tallest railroad bridge in the world and remains today the tallest in the United States. A viaduct is a type of bridge that spans a greater or taller distance than typical bridges and is designed to carry immense weight, such as trains loaded with freight. (MG 183-F14-008.)

Many of the bridges built for the Camas Prairie Railroad could be moved in order to allow the passage of boats. Unlike a drawbridge that angles upward, bridges like this one near Lewiston were swing bridges with a section that could turn horizontally like a doorway to allow the passage of boats. (PG 90 1-2-68-1012.)

This train is headed to Lewiston, pulling a US Mail Railway Post Office (RPO) car. These rolling post offices were typically included on passenger service routes and were designed so that postal employees could sort mail as they traveled, speeding up delivery. Mail was carried by railway for almost 90 years, until around 1970. (MG 183-F13-012.)

On the camas prairie, maintaining safe lines was an ongoing challenge. This image shows the effect of erosion on tracks laid near the water. This Northern Pacific passenger train had an unexpected detour into the Clearwater River near Arrow Junction in 1936. (PG 90-04-081.)

One of the challenges of building a railroad in a dynamic landscape is that while train tracks may seem to be permanent fixtures, the ground underneath them can swell, shrink, harden, and soften, and when a locomotive slips the tracks, the results can be dramatic. Here, locomotive No. 2211 had hardly left town before succumbing to unfavorable track conditions. (MG 183-F16-055.)

A rock fall caused the derailment of Union Pacific train No. 860 on the Camas Prairie Railroad 17 miles west of Lewiston, resulting in engine No. 2881 lying on the bank. This passenger train ran between Lewiston and Ayer Junction and was equipped with acknowledging devices and a primitive automatic train control function that allowed for safe speeds. Sadly, engineer Jack Wilburn was severely burned in this accident. (MG 183-F09-006.)

This 1951 photograph offers a striking view of Northern Pacific locomotive No. 1506 pulling train Extra 1506 east near Nucrag, Idaho. The dramatic swells were caused by glacial movement during the last ice age and a series of small faults just beneath the surface. (MG 183-F19-016.)

The Union Station depot in Moscow stood from 1936 to 1965 just south of Sixth Street and was straddled by lines for the Union Pacific and Northern Pacific. To the east ran the UP branch to Pullman, and to the west ran the NP branch that connected with the main line of the Camas Prairie Railroad. (PG 9-17-12a.)

The town of Reubens was the first stop heading east toward Grangeville after the harrowing tunnels and bridges of Lapwai Canyon. No less than nine routes were surveyed between Reubens and the next stop to the west, Culdesac, on what is part of the steepest section of the entire line, with a 3 percent gradient. (PG 38-1396.)

This 1927 image shows the construction of a trestle near Reeds Creek divide near Headquarters, with staggered large timbers and temporary braces providing the strength needed to hold up locomotives loaded with timber. Two men balance on the supports, inspecting the work. (PG 13-6034.)

The Camas Prairie Railroad connected the forests of the inland northwest to facilitate the transportation of lumber to the rivers and major train lines that would carry the wood east. At one point, almost all of the white pine processed in the United States traveled via this modest railway. (PG 13-535.)

The impetus for the creation of the Camas Prairie Railroad was for it to serve in the place of rivers, which were used elsewhere to move lumber from forests to the cities that could put it to use. Here, newly fallen logs are on their way to Lewiston to be floated on the Snake River and sent to their ultimate destination. (MG 183-F12-006.)

Right around the time that the Camas Prairie Railroad was expanding to the heavily forested areas to the east, the Clearwater Timber Company built an extensive mill in Lewiston. Here, a crane loads gravel into a railcar to help with construction. The massive structure later became a dry shed, where lumber is laid out to dry to make it suitable for use as a building material. (MG 457-202-15-034.)

Even if logs traveled by rail to get to the mill in Lewiston, they ended up in a storage pond. The pond was adjacent to the Clearwater River but not directly connected, to avoid seasonal flooding. Ponds helped mill workers by allowing them to easily move and position the buoyant logs for processing. (MG 457-202-12-009.)

It took 1,500 men just over a year to complete the 41 miles of Camas Prairie Railroad track from Orofino to Headquarters, the fourth and final subdivision of the line. These proud men, who are smiling and showing off their hand tools, must have just completed their section. The long-handled spike mauls many of them are holding were used to drive spikes into the wooden ties to hold the tracks securely in place. (MG 457-163-329-001.)

Simple tools were frequently all workers needed to keep the trains running. Because it was much cheaper, coal had largely replaced wood as the fuel of choice for steam engines by the start of the 20th century. The amount of coal that needed to be shoveled varied based on the weight of the train, the incline of the track, and so forth, but an average baseline was about 15 pounds of coal per mile, or about four shovelfuls. (PG 91-630.)

This Shay locomotive stopped near Potlatch, which allowed the crew to pose on or around the train and show off the extensive load of lumber it was hauling. The funneled smokestack was a signature of wood-burning steam engines like the Shay. Although coal was more efficient, wood-burning engines had advantages in areas where logging was taking place. (PG 6-62-2.)

The first log train to roll out of Jaype (pronounced "jay-pee"), one of the larger millsites along the fourth subdivision of the Camas Prairie Railroad, was an ambitious one. The weights and lengths of the trains on the entire line were closely monitored in an attempt to avoid accidents caused by the steep grade, tight curves, and spindly trestles. (MG 457-204-70.)

Lake Chelan
Chelan
Columbia
N
T
O
N
SPOKANE
Mansfield
Waterville
Withrow
Coulee City
Douglas
McCue
Palisades
Almira
Govan
Wilbur
Creston
Hartline
Coulee Jct
Bacon
Davenport
Fellows
W SPOKANE
COWLES
MARSHALL
Dennys
Bluestem
CHENEY
GEIB
MASON
CROSKEY
WELLS
Morocco
Downs
Lamona
Irby
Krupp
Wilson Cr.
Adrian
Odessa
Marcellus
Euphrata
Naylor
Winchester
Quincy
Trinidad
Columbia River
G. N. RY
Leavenworth
Peshastin
Monitor
Malaga
Vulcan
Moody
Schoonover
Ritzville
Ruff
Tiflis
Schrag
Bassett Jct
Warden
Othello
Roxboro
PALM LAKE
ASHBY
EMDEN
TESKE
MARENGO
THAVIS
MACK
ANKENY
HOOPER JCT.
Murdock
Ellensburg
Kittitas
Boylston
Jericho
Smyrna
Corfu
Anson
Beverly Jct.
Columbia River
YAKIMA
UNION GAP
PARKER
DONALD
SAWYER
BUENA
ZILLAH
GRANGER
EMERALD
SUNNYSIDE
MIDVALE
GRANDVIEW
CAPP
NORTH PROSSER
Toppenish
Alfalfa
Satus
Mabton
Byron
Prosser
CONNELL
Mesa
WASHTUCNA
HOOPER
GORDON
PAMPA
WINONA
SUTTON
LA CROSSE
ROCKFORD
DARKNELL
FAIRFIELD
LATAH
FLETCHER
OAKESDALE
ENDICOTT
THERA
DIAMOND
MOCKONEMA
ELBERTON
COLFAX
EAST SPOKANE
DISHMAN
CHESTER
MICA
FREEMAN
PARK
MAGALLON
RUXBY
MATTHEW
SCOTT
WALKER
WALKER PIT
SHEFFLER
SIMMONS
PAGE
RIPARIA
TUCANNON
STARBUCK
RELIEF
ALTO
MENOKEN
PRESCOTT
ENNIS
DELANEY
JACKSON
POMEROY
ZUMWALT
HOUSER
TURNER
LEWISTON
Camas Pr.
KENNEWICK
LESLIE
HEDGES
ATTALIA
WALLULA
COLD SPRINGS
RIVERVIEW
JUNIPER
SAND
UMATILLA
BAILEY
IRRIGON
JUDSON
TOUCHET
LOWDEN
WHITMAN
ARTESIA
AUKER
BOLLES
RUSSELL
VALLEY GROVE
HADLEY
WALLA WALLA
Blue Mts.
Grande Ronde River
SPOFFORD
MILTON
PRUNEDALE
BLUE MOUNTAIN
DOWNING
WESTON
ATHENA
ADAMS
BLAKELEY
HAVANA
SAXE
PENDLETON
HERMISTON
PENDAIR
ECHO
NOLIN
YOAKUM
CAMPBELL
BARNHARDT
RIETH
SPARKS
LENS
PILOT ROCK
HINKLE
STANFIELD
CLARKE
MUNLEY
ORDNANCE
WESTLAND
BOARDMAN
CASTLE
HEPPNER JCT.
WILLOWS
SILICA
ARLINGTON
GILMORE
BLALOCK
RAMSAY
WASCO
KLONDIKE
SANDON
HAY CANYON
CECIL
MORGAN
IONE
JORDAN
LEXINGTON
HEPPNER
SHUTLER
MCNAB
ROCK CREEK
BARNETT
MIKKALO
CLEM
SPEECE
GWENDOLEN
CONDON
GRASS VALLEY
BOURBON
KENT
GIBBON
BONIFER
SLOAN
DUNCAN
NO. FORK
CAMP
HURON
MEACHAM
KAMELA
HILGARD
PERRY
LA GRANDE
LONE TREE
HOT LAKE
UNION JCT.
PYLE
LOOKING GLASS
RONDOWA
VINCENT
MINAM
SEVIER
WALLOWA
LOSTINE
ENTERP.
JOSEPH
ELGIN
IMBLER
ALICEL
CONLEY
ISLAND CITY
Cove
Union
CROOKS
TELOCASET
SAGO
LUN
NORTH POWDER
HUTCHINSON
HAINES
WING
Blue Mts.
Ukiah
Parkers Mill
Fossil
Winlock
Antelope
John Day

This tangled web of a map shows the difficulties presented by the terrain when companies were trying to connect locations by rail. Small homegrown businesses competed with national empires to carry to larger markets not just people and equipment but goods and commodities produced in the fields, forests, and mines of the region. (PG 90-3-9-78-002.)

Building a railroad is a long and grueling process. The ground must first be prepared by cutting through hillsides to make tunnels or making wedge-shaped cuts and filling in low points or building bridges over the larger gaps. The elevation of the fourth subdivision of the Camas Prairie Railroad rises about 2,500 feet from Orofino to Summit, then drops down a steep 2.2 percent grade into Headquarters. There was no getting around it—human-, horse-, and machine-powered labor were all required in order to complete the railroad. (MG 457-163-329-002.)

Two

A Railroad on Stilts

The construction of the Camas Prairie Railroad was unprecedented, with a remarkable number of tunnels, steep grades, and high trestles along a relatively short route. Given the variety of terrain, ecosystems, and goods and businesses along the route, the line was built in four subdivisions. The first traveled from Lewiston southeast to Stites, the second southeast from Lewiston to Grangeville, the third from Lewiston west to Riparia, and the fourth branching off the first line at Orofino to head northeast to the forested area of Headquarters.

The CPR began by primarily focusing on the overland shipment of agricultural products produced in the fertile rolling hills of the camas prairie that could be accessed by the first and second subdivisions. The region had already been established as an exceptional producer of wheat and other dryland crops on the hills, and fruit in the lower valleys, but the only way to ship the produce to larger markets involved steamships leaving port in Lewiston. Nearby heavily forested areas enticed large companies like Weyerhaeuser and Potlatch Forest Services to partner with the railroad to cut, process, and haul lumber. When the Clearwater Mill was built in Lewiston in 1926, it was the largest white pine mill in the world.

The lure of extractive resources and abundant agricultural products spurred two smaller railroad companies—the Nez Perce & Idaho Railway (NP&I) and Craig Mountain Railroad—that connected directly to the second subdivision. It is notable that the CPR itself owned no locomotive equipment, unlike virtually every other railroad system in the country, but rather rented equipment through cooperative agreements with its Union Pacific and Northern Pacific parent companies. Governance was also shared, as the president's position rotated every year between NP and UP general managers. The CPR spanned the era of both steam and diesel trains, as well as the time when the telegraph system progressed from Morse code to computers. For a brief time, small electric trains (colloquially known as doodlebugs) were used to cart passengers across the network.

Many people anxiously anticipated the arrival of locomotives on the rural prairie, and the construction drew plenty of interest. An article in the March 28, 1908, *Camas Prairie Chronicle* stated that two women got into a construction bucket suspended on a cable used to transport fill material and concrete for the supports (possibly like the ones in this photograph) under the Lawyers Canyon viaduct and were hauled halfway across the span to enjoy the view. (PG 5-093-2b.)

Originally known as Arrow Junction, this was where the Northern Pacific Railroad met up with the Camas Prairie Railroad, as well as where Potlatch Creek flowed into the Clearwater River; this convergence is now known simply as Arrow. The small community was named for the large number of arrowheads found when the Northern Pacific excavated near the banks of Potlatch Creek. (PG 99-030-061.)

Sometimes the mountains were just too steep, and switchbacks or trestles would not work, which forced the railroad to go through the mountain either by tunneling through it or flattening a section. Excavating through mountains was a slow process. This path between Orofino and Headquarters used a rail line temporarily laid on cut logs to help remove debris. (PG 13-6033.)

Two workers sit on a plank dangling off the edge of the Lawyers Canyon viaduct. This is one of only a few steel bridges in the Camas Prairie Railroad system and, at approximately 280 feet, the tallest bridge on the line. When it was completed in August 1908, it was supposedly the highest railroad bridge in the world at that time. (MG 183-F16-010.)

This is one of the switchbacks on the steep grade between Reubens and Culdesac. The trestle at the lower level in the distance is only a few minutes away via train. The line twists and turns in order to move people and goods safely up and over the steep, narrow canyons. (PG 16-1-23.)

Train No. 857 winds its way out of tunnel No. 1 at milepost 20 between Culdesac and Reubens. Route 95 winds below at a cross angle, showing the different relationships that rails and roads have with this rugged landscape. Tunnel No. 1, also known as the Horseshoe Tunnel, makes an 882-foot-long hairpin curve through the mountain. (MG 183-F24-008.)

This train is pushing its way through the snow at Nucrag in Lapwai Canyon on the second subdivision, about halfway between Lewiston and Grangeville. Winters could be quite harsh, with an average yearly snowfall around 25 inches. Nucrag now only exists as a historical footnote. (MG 183-F16-033.)

Employees proudly pose with Nez Perce & Idaho Railroad locomotive No. 4. in Craigmont. In the doorway, a cab curtain is draped across one of the men. During inclement weather, the crew might hang canvas curtains in an effort to keep warmth in—and cold out of—the cab. (MG 183-F25-036.)

The old steam locomotive No. 4, formerly Northern Pacific's No. 684, is shown just ahead of the caboose being pulled over a wooden trestle near the town of Nucrag. Crews were apprehensive about No. 684 remaining on the rails because of its old tires and sharp flanges. The metal tires were subject to fatigue and cracking after years of use. The severe temperature swings they faced when being changed, such as the extremely high heat used to mount them on the wheels and then being securely fastened by cooler temperatures, could also result in serious damage. Nucrag was created to support newly developed settlement along the line and was named after two Camas Prairie Railroad employees—Mr. Newton, a conductor, and Mr. Craig, an engineer. (PG 16-1-21b.)

Riparia was a regional hub for the Camas Prairie Railroad, the Oregon Railroad and Navigation Company, and the Snake River Valley Railroad. The CPR line between Riparia and Lewiston was the first significant railway in the area and began operations in 1908. It eventually connected all the way through to Grangeville at the far terminus of the Camas Prairie Railroad. (PG 99-R-017-006.)

Earl Cash worked for the Camas Prairie Railroad for 47 years, starting in June 1920. A diligent and hardworking employee, Cash quickly rose to become foreman of bridge construction and maintenance and not only kept up repairs and maintenance of the entire railroad but helped to build the line from Orofino to Headquarters in 1925 and 1926. (MG 183-F16-007.)

The agricultural and forest potential of the camas prairie region spawned two smaller branch railroads off the Camas Prairie Railroad—the Nez Perce & Idaho Railroad and the Craig Mountain Railroad. Just 14 miles long, the NP&I traveled from Craigmont on the CPR line out to the small town of Nez Perce, where the railroad's main offices and a depot were also located. (MG 183-F25-012.)

The climate of the Pacific Northwest lends itself to dense, wet snow that could sometimes be too much for a locomotive. Due to the extensive manual labor required, some legs of the Camas Prairie Railroad would only be traveled every few weeks during the winter, making the trains a very welcome sight to the residents of small towns on the rolling prairie. (MG 183-F14-013.)

Northern Pacific Railroad locomotive No. 1521 travels on a flat section near Reubens. The cars trail off into the distance with engine No. 684 in tow. There are many variables to consider, but in general, a steam engine could pull roughly 20 cars at a full running speed of 50 miles per hour. (MG 183-F25-024.)

Maintenance of the line includes removing weeds and debris to keep them from encroaching on the track. Here, J. Hays and a Camas Prairie Railroad work crew use a mobile weed-burner to remove weeds, with a motor car hauling barrels of water and a sprayer following close behind to put out the fires. (MG 183-F16-006.)

All three locomotives of the Nez Perce & Idaho Railroad are captured in this one image. The annotation states that these 44-ton, 38-horsepower General Electric locomotives were painted a bright red and orange, but since the original photograph is black and white, the vivid colors must be imagined. Electric locomotives offered railroads an economical way to haul light loads and passengers. (MG 183-F11-003.)

The joint nature of the Camas Prairie Railroad is evident in this photograph showing both Northern Pacific and Union Pacific branded engines pulling a train east over the Clearwater River toward Lewiston and probably on to Riparia, where it will connect with other railroads to move lumber and other products to more distant markets. (MG 183-F19-033.)

A woman is waiting on the platform at the Lewiston depot. The Camas Prairie Railroad ran passenger trains from Spokane to Grangeville for decades. However, as personal automobiles became more common and paved roads began to connect the cities of the camas prairie and the Palouse, passenger service on the railroad fell out of fashion and eventually ceased. The last regularly scheduled passenger train left Grangeville on August 24, 1955, and larger stations such as Lewiston or Orofino, which connected passengers through Northern Pacific lines to Spokane had completely discontinued passenger services by 1965. (PG 91-627.)

Northern Pacific locomotive No. 1506 takes on sand at the coal dock in the East Lewiston rail yard. All trains heading up and into the canyons carried sand to improve traction. The sharp turns, spindly trestles, and steep gradients required trains to slow down to a point where the wheels would slip on the tracks, and sand would help prevent this. Specially designed equipment called a sandbox delivered the sand directly to the rails using gravity, steam, or compressed air. (MG 183-F19-011.)

Union Pacific No. 2037 and Burlington Northern Nos. 2076 and 2077 are working together to get across a trestle at Lapwai Canyon. One of the most dramatic sections of the railroad, the second subdivision includes not only several tunnels and tall trestles within a few miles but also a steep grade that often required multiple engines to pull the heavy trains. (MG 183-F26-031.)

This image offers a glimpse into the dark depths of tunnel No. 7 as the tracks head east toward Grangeville. This is the last tunnel in the harrowing Lapwai Canyon section. This concrete-and-timber-lined engineering feat curves 529 feet through the hillside just before Reubens. The exterior structure helped to stabilize the opening and direct snow and debris away from the exposed track. (MG 183-F14-028.)

Smoke is seen rising from the tracks as the brakes heat up from constant deceleration on the way through Lapwai Canyon back toward Lewiston. The railroad anticipated overheating and put a requirement in the timetables for engineers to stop and wait 10 to 15 minutes at multiple stations in the event the wheels got too hot. Limits were also placed on how much weight each train could pull. (MG 183-F19-017.)

At the top of the camas prairie, one would never know this railroad gained the nickname "Railroad on Stilts" due to the numerous spindly wooden trestles over which the trains had to travel. The rugged hills in the distance, however, betray the hope of maintaining the gentle slopes of open farmland. (MG 183-F28-004.)

Shown here just ahead of the caboose, Nez Perce & Idaho Railroad locomotive No. 4 is being towed from Culdesac down to Lewiston. This photograph was taken near Lapwai. Due to the complicated braking and transmission systems of locomotives, towing involves significant mechanical work in order to make everything safe for transport. (MG 183-F25-027.)

Oil-burning steam trains were used to haul heavy loads of lumber out of the forested areas to reduce the risk of starting a wildfire. Coal-burning engines can emit embers in their smoke and exhaust, which can ignite brush and trees near the tracks. (MG 183-F19-028.)

In the early days of rail travel, turntables provided one of only a few ways a train could turn around and were invaluable for locomotives that needed to change direction. This image shows a bit of a traffic jam, with Union Pacific engine No. 2100 and Northern Pacific engine No. 1735 waiting their turns as Union Pacific No. 3217 goes ahead in East Lewiston. (MG 183-F19-009.)

Train No. 622's conductor takes a moment to telephone the dispatcher during a quick stop at North Lapwai, Idaho. The city of Lapwai was the second-northernmost stop along the second subdivision of the Camas Prairie Railroad. Founded as a Protestant mission in 1836, the spot became Fort Lapwai in 1863 and was incorporated as a city in 1911, just a few years after the Camas Prairie Railroad went into operation in 1908. (MG 183-F19-007.)

Nez Perce & Idado train No. 9 is pictured in Craigmont in 1951. It had just completed a midnight run from Nez Perce without any oil in the journals. A notation on the reverse observes, "A wonder they made it." (MG 183-F25-020.)

Train No. 885 is crossing the east leg of the wye switch near Spalding. A wye is a triangular junction with a railroad switch that allows trains to turn onto a new track or, with some maneuvering, turn around. (MG 183-F24-007.)

A Lewiston Lions Club excursion train is pictured in 1960 on the trestle near Reubens. This excursion took place after passenger service on the second subdivision of the Camas Prairie Railroad ceased. It was impossible for the railroad to compete with buses that could now carry passengers on the nearby highway. However, the promoters felt the scenery on the line was well worth the trip. (MG 183-F13-003.)

I.A. Wolters, general manager of the Camas Prairie Railroad, is inspecting locomotive No. 4 from the Nez Perce & Idaho Railroad before it moves over to the CPR line. Wolters, formerly of the Northern Pacific, took over management of the CPR in January 1953 and served in that position until his retirement in January 1957. (PG 16-1-20.)

When freight trains carry heavy loads, a second locomotive is sometimes attached at the rear. A rear locomotive reduces the amount of torque that must be applied by the lead engine, preventing pivoting, which is when the front wheels of a locomotive lift up off the track. The push from the back also reduces stress on the hook-and-knuckle joints that connect the cars and prevents runaway cars from moving downhill should one of the joints break. (MG 183-F12-008.)

Railroad speeders were used to quickly move crews or small amounts of supplies from one station to another. Originally gas-powered railcars were used, but were replaced by specially adapted trucks such as the one seen here being loaded with supplies in Headquarters to take out to the logging camps that were accessible only by spur tracks. (MG 183-F12-007.)

This close-up of felled trees hauled by the Camas Prairie Railroad shows rotten logs mixed in with viable timber, with all of the wood headed to the same mill. Wooden supports hold the logs in place on open railcars. Lumber played a dominant role in the establishment of the CPR. (PG 13-4975.)

The remote locations of logging operations were typically only accessible by railroad, and the short amount of time workers spent extracting the timber meant that they were constantly on the move. Bunkhouses and such were built on railcars so they could be easily moved. (MG 457-161-234-003.)

Rolling villages traveled along spur routes into the forested areas so that loggers and support staff could have a place to call home. Union efforts ensured that living quarters were heated and periodically cleaned. Temporary steps and awnings were attached to entrances but always ready to be taken down and moved to the next location. (MG 457-161-234-001.)

This 1940s logging crew is loaded on a speeder and ready to head out of Headquarters on a Camas Prairie Railroad spur track. The one woman on board was likely employed as a camp flunky—a person who worked in the cookhouse, helping to prepare and clean up after meals and ensure the men heading into the woods had a packed lunch to take with them. (MG 457-161-234-002.)

In addition to utilizing circular snowblower-type plows, trains also used cars with wedge-shaped front ends to push snow off the tracks. Large, curved pieces of metal also helped push snow from the edges of the track to keep it from becoming unmanageable and falling back down in the path. Deep snow becomes practically impassable as the mound grows taller than the plow and builds up on the track under where the plow stops in between the rails. (Above, MG 457-164-491; below, MG 183-F26-024.)

The spray of material being kicked up by this train is ballast, which is typically a gravel or small-rock substrate that is used in building and repairing the tracks. The particulate matter is packed down around the ties and acts as a solid support system that keeps everything in place yet is still permeable enough to let rain and snow drain through it. (MG 183-F14-029.)

Regardless of whether this train was powered by coal or diesel, a large plume is being emitted from the smokestack as it gains power and speed. The characteristic bell shape of the smokestack on this Northern Pacific locomotive leaving the East Lewiston Yard in 1950 shows it was a steam engine. The bell chimney allowed the smoke or steam to billow up and away from the engine, providing better vision for the engineer. The confined spaces of the long, curving tunnels on the Camas Prairie Railroad and the amount of exhaust given off inside created hazardous conditions for anyone on the train. (MG 183-F19-006.)

Disaster could strike at any time. In the above image, flooding has washed out a bridge near Kamiah. This route closely followed the south fork of the Clearwater River, which experienced one of its highest floods on record in the spring of 1948. The accident in the undated photograph below took place on a trestle just outside tunnel No. 1, the Horseshoe Tunnel. This bridge was eventually filled in and replaced by a metal culvert. (Above, PG 13-65; below, MG 183-F27-002.)

The large cylindrical object on this flat railcar parked outside the Lewiston Mill is a Yankee dryer. Typically made of cast iron, these cylinders use pressure created by injecting steam into and throughout the center to remove excess water from wood pulp that will be used to make paper. The brittleness of cast iron made it sensitive to temperature fluctuations, and occasionally, a Yankee dryer would explode if not handled properly. For this reason, modern versions are made with steel. (MG 457-165-899.)

Each route of the Camas Prairie Railroad closely followed a river or creek. Here, a Grangeville-bound train leaves Lewiston along the Clearwater River, pulling extra empty cars in order to have plenty on hand to haul the harvest back down and on to larger markets. (MG 183-F24-009.)

This is an early example of a loader—essentially a small crane on a railcar bed that is used to move loads on and off of trains. In addition to being run on the track, some loaders were also dragged alongside the track with metal cables so they could access different cars without coupling and decoupling individual cars. (MG 183-F12-011.)

To keep railcars in place while they were being loaded, workers set a hand brake by manually turning a wheel on the end of the car. In some trains, this will activate the braking system on all the cars, but on others it will only engage one car's brakes. For safety, when the train is moving, either an electronic or air braking system is used. (MG 457-165-1081.)

With an established end-of-the-line stop, Headquarters transitioned from a temporary logging camp to an actual town. A post office was in place by 1928, along with permanent buildings such as schools, stores, and even a swimming pool for use in the summer. At its peak in the 1960s, Headquarters was home to 300 people and supported 15 logging camps. (MG 457-165-976.)

Railroad camps could be desolate places. Constantly moving from place to place to extract all the timber was grueling work. Two men in the distance are holding horseshoes and possibly attempting to start up a game to while away their free time. (MG 457-163-466.)

A crew relaxes alongside a train parked on a siding at Haley. Sidings, auxiliary tracks adjacent to the main line, were built so that trains could pull off the tracks to allow other trains to pass. Once a small logging community, Haley no longer exists. (MG 183-F16-019.)

An engine crew stands at the ready outside the Lewiston depot. If there was any doubt after noting the large shovel, the headlight resting on top of the engine also suggests this locomotive was among the first generation of steam engines driven by coal. Smokestacks shrunk and rounded boilers became more prevalent as railroads moved away from using wood as fuel. A coal car is attached behind the engine. (PG 91-631.)

Three

An Abandoned Track Rediscovered

Despite steady rates or even increases in shippable products, the Camas Prairie Railroad began abandoning less-traveled lines in 1985. As the forests became depleted, improvements to mills and wood-processing facilities and the never-ending need for building supplies forced lumber companies to adjust how they operated. In the 1950s, the Lewiston Mill switched from lumber products to paper, converting sawmill waste into household paper products so the company would not have to rely so heavily on harvesting timber.

Changing markets and diseases in wheat caused the agricultural benches of the camas prairie to branch out from wheat to alfalfa, barley, and oats. Increases in commodity production, along with the desire to ship smaller quantities at the farmers' convenience, brought about improvements to other forms of transportation. US Highway 195, which connects north and south Idaho, closely follows the route of the Grangeville branch of the CPR.

The decline of the CPR continued until 1997, when it was sold to North American Railnet, a Texas-based speculation firm specializing in short lines. Delays in attempting to get the trains running on a consistent schedule frustrated farmers who worried about getting perishable crops to market in a timely fashion. Despite the promising new ownership, the railroad continued to be abandoned piece by piece until it ceased operation in 2000. The CPR's trestles and tunnels still dot the landscape, but the tracks on most of the rural sections of the line have been removed.

In 2002, the remaining portions were purchased by Bountiful Grain & Craig Mountain Railroad, which sold it to Watco in 2004; Watco renamed it the Great Northwest Railroad and began leasing track access to other railways. Further improvements to the Lewiston Paper Mill and agricultural cooperatives that can put together large quantities of products to ship from one location ensure continued demand for high-volume transportation. The Great Northwest Railroad is once again in operation with shipments of wood and food products.

Northern Pacific's engine B-14 stopped in Lewiston to unload the mail and express after returning from Grangeville. When this photograph was taken on June 23, 1954, trains left Grangeville each day at 11:05 a.m. to make the almost three-hour trip to Lewiston, arriving at 1:55 p.m. and stopping at no fewer than 18 stations along the way. (MG 183-F13-008.)

Designed in 1908, the Camas Prairie Railroad depot in Lewiston is one of the grander depots in northern Idaho. The structure includes a large general waiting room, separate lounge areas for men and women, a ticket office, and baggage storage areas on the first floor. Since Lewiston was also the headquarters of the CPR, offices for the roadmaster, superintendent, and support staff were located on the second floor. In 1973, the Lewiston depot was listed in the National Register of Historic Places. (MG 183-F26-020.)

At the end of a statewide tour in 1951, the Idaho Land Board took the opportunity to pose for a group photograph near the end of a spur line. Close attention had to be paid by anyone traveling up this short section of track that abruptly ends in a remote forested area so the train would not fall off the end of the line. (PG 13-5193.)

All the tracks are full of trains waiting to be loaded or unloaded on a busy day in 1971 at the Cottonwood siding. Locomotives of this era were diesel-powered rather than steam-powered. The diesel engine drives an electric generator that in turn powers the locomotive's motors. (MG 183-F14-031.)

Five years after the discontinuation of passenger service on the second subdivision of the Camas Prairie Railroad, the Lewiston Lions Club sponsored an excursion train from Lewiston to Grangeville in 1960. Two diesel engines, eight passenger cars, and one baggage and refreshment car carried anyone willing to pay the $5 fare to travel along the scenic "railroad on stilts." (MG 183-F13-002.)

Seen here traveling west through the gentle rolling hills of the camas prairie near Craigmont is the Extra 1521 hauling a load of freight. Before technology allowed for communication between trains, the train schedule was strictly adhered to so as to avoid collisions. Regularly scheduled passenger trains had the highest priority, followed by several classes of trains below them, with the lowest being "extra" trains, which were unscheduled and had to cede right of way to every other train on the track. (MG 183-F19-022.)

Even on a seemingly stable and flat section of track, environmental damage can occur. Sudden and extreme changes in the winter of 1968, during which temperatures ranged from -10 to 30 degrees Fahrenheit in one week and from 30 to 60 degrees Fahrenheit the next week, along with higher than normal precipitation rates in the inland northwest region, caused extensive flooding. In East Lewiston, the tracks buckled as the ground underneath them was washed away in December 1964. (MG 183-F14-002.)

The harvesting of timber frequently left behind many waste products, such as small branches and brush called slash that could not be used for lumber. The conical structure at left in this photograph is a slash burner, which were later banned due to the amount of air pollution they created. They had no filter, only a metal screen at the top to catch sparks or embers. (MG 183-F28-001.)

The Northern Pacific gas/electric B-12 engine is shown passing another locomotive waiting on a siding at North Lapwai. These smaller vehicles came into use as the highway system grew after World War II and were in service until August 1955. (MG 183-F26-026.)

Prior to the completion of Dworshak Dam on the North Fork of the Clearwater River in 1973, the town of Orofino frequently flooded. The US Army Corps of Engineers, which built and maintains the dam, estimates that the dam has prevented approximately $2.8 million in damage since its construction. (PG 13-5104.)

Although it may look reminiscent of something in a horror movie, the rotary snowplow was an invaluable tool for keeping tracks clear in the winter. The rotating blades pick up snow and throw it up and over the bank, allowing for rotaries to be used in deeper drifts and where the tracks are laid in steep cuts in the earth. (MG 183-F26-007.)

Snow led to frequent battles on the windswept prairie. It was not uncommon for drifts to pile up higher than the trains, meaning that plows would have to slowly chip away at the frozen mass. High winds would also pick up dirt and sand from uncovered fields and deposit it between the layers of snow—one engineer stated that it looked like they were cutting into a marbled chocolate cake. (MG 183-F16-036.)

An excursion train travels alongside a creek between Culdesac and Reubens. Excursion trains are specially chartered, perhaps by tourists or to convey large numbers of people to an event. It is not unusual for an excursion train to make a round trip back to its point of origin. (MG 183-F12-003.)

Train No. 314, behind Northern Pacific No. 1365, is passing Union Pacific No. 2100 and Northern Pacific No. 1521 at Spalding. Spalding is 10 miles east of Lewiston, the central hub of the Camas Prairie Railroad, which sits on the border of Idaho and Washington. (MG 183-F19-001.)

The 563-foot-long Camas Prairie Railroad tunnel No. 3, the second-longest tunnel on the Grangeville branch, was plagued with difficulties such as frequent cave-ins and rockfalls. Daylight was visible at the other end, giving a false sense of security to this tangent (straight) section of track. (MG 183-F14-027.)

Before the damming of the Clearwater River, timber was floated downstream from remote logging areas in controlled batches called log drives. When the logs become stuck in shallow areas or entangled with each other and debris, as shown here, they are called logjams and can be quite dangerous. The train in the distance is carrying neatly stacked lumber to the mill in Lewiston. (PG 99-R-012-10.)

The disastrous results of a derailment at milepost 2.5 near Lapwai are shown in this photograph taken in 1975. Train derailments were common until about 1980, when a combination of improved materials and track replacement (due to age) helped to create a safer environment. (MG 183-F14-006.)

Along with having frequent maintenance issues and damage caused by flooding, bridge No. 50.1 over the Clearwater River, on the first subdivision, was also not immune to fire. Originally constructed in 1881 for a completely different river crossing, this bridge was installed near Kamiah in 1900 after others attempted to use it in at least two other places. (MG 183-F14-014.)

The Camas Prairie Railroad's original wooden trestles were built with locally sourced pine that unfortunately deteriorated very quickly. Cost estimates determined that replacing all of the trestles with steel would be too expensive, so the bridges underwent continual maintenance and monitoring for signs of damage. Individual pieces that were found to be lacking were replaced with treated lumber. (MG 183-F16-001.)

This dramatic view of the rough terrain that had to be conquered during the construction of the Camas Prairie Railroad hints at the possible instability of the line. The blasting debris from the creation of a tunnel and the manufacturing of a flat bed of earth for the tracks was used to shore up the bank where the tracks were laid. (MG 183-F13-011.)

In the 1960s, changes were made to the route built by the Union Pacific from Riparia to Lewiston. Two dams constructed on this small section of the Snake River caused the water to rise, flooding the original line and stations along the way. The economic viability of agriculture on the camas prairie ensured the relocation of the route. (MG 183-F20-003.)

Looking up from beneath is one way to appreciate these majestic elevated tracks. Steel beams are favored for use in train lines because of steel's ductile strength. Trains generate a remarkable amount of vibration when they pass over tracks, and steel has a high level of stamina for this type of stress. Steel beams are also quite efficient at distributing weight across a framework. (PG 99-R-686-004.)

Constructed in 1927, the north-south highway (now US Route 95) has been improved and expanded, requiring the reconfiguration of multiple bridges on the second subdivision between Lewiston and Grangeville. Including old and alternate routes, the Camas Prairie Railroad crossed the highway six times in 66 miles. Improvements made to the highway in the 1950s became a major factor in the discontinuation of passenger service on the railroad. (MA 2016-53-008.)

These two boys spent a pleasant afternoon under one of the many crossings the Camas Prairie Railroad made as it wound its way along Orofino Creek. Orofino—which means "fine gold" in Spanish—got its name from the abundance of gold mines in the area. Orofino Creek is also a popular spot for fishing. (PG 13-7289.)

The Camas Prairie Railroad boasts the tallest trestles in the world, with some standing a remarkable 250 feet above the ground. These towering constructions were made using a combination of locally logged heavy timber and structural steel. The branch from Lewiston to Grangeville is the most visible and what gave the CPR its nickname, but there are just as many magnificently

curving trestles between Orofino and Headquarters. This last branch traveled through a heavily forested and sparsely populated area, so its trestles are hidden away in the woods. It would be a feat to construct these trestles today with modern machinery, but they were built by hand over a century ago! (MG 183-F12-010.)

On a dreary winter day, the view from the vantage point next to the tracks offers an impression of just how steep a trip the railroad had to make as it traveled up from the valley. The bridge in the distance is only a few miles away by train. (MA 2016-53-001.)

With the rising costs of offering full passenger service trains or even combination passenger/freight lines, the Camas Prairie Railroad moved passenger service, mail, express, and other small deliveries to two gas/electric motors in the 1940s to compete with passenger buses and delivery trucks. Here, they are heading out of Lewiston, with one going to the far reaches of Grangeville and the other to Stites. (MG 183-F19-014.)

Whether they are in the heart of a town or the outskirts like this one in Winchester, Idaho, railroad depots are a vital part of the rail system. Depots provide regularly scheduled stops, allowing passengers and companies to conduct business and travel with ease, lending stability to a rural way of life. (PG 6-62-9.)

The importance of maintaining timbers was never more important than on this fateful day in 1944, when one section gave way and a whole car dangled in the air by one cable. The elevated tracks led to the coal dock, which had a height offering easier access to load coal and other supplies. (MG 183-F16-058.)

This engine on the Headquarters branch made it about halfway to its destination before hitting a soft spot on the tracks and sliding down an embankment near Poorman, Idaho. Both the engineer and fireman were injured, and given where the engine landed, there was no hope of returning it to the tracks, so it was cut up for scrap metal. (MG 183-F09-013.)

These two engines, likely pulling a heavy load of lumber, are stopped on one of the many bridges over Orofino Creek on their way back to Lewiston. Maintenance was a never-ending process, as demonstrated by the man kneeling by the track next to the second engine and checking the equipment. (MG 457-163-376.)

This quiet, idyllic summer day belies the traffic of the Camas Prairie Railroad line to Headquarters. For many years, trains ran from Orofino to Headquarters six days a week. To keep up with taking the lumber to connecting trains in Lewiston and replenishing the line with empty cars, a second train called the Night Logger ran overnight from Lewiston to Orofino six nights a week. (MG 457-164-740.)

Union Pacific locomotive No. 246 and train No. 858 are pictured at the lumber mill siding at North Lapwai. If a siding is connected to a main track on both ends, it is called a loop, dead-end, or stub. (MG 183-F28-006.)

A light dusting of snow is no match for a lumbering locomotive. However, the unseen dangers that may lurk around the corner could give even the most intrepid engineers pause. Snow drifts, flooding, washouts, and rockfalls were common. To keep train operators alert, the use of alcohol was strictly prohibited. As with many railroads, the Camas Prairie had its own police to contend with any lawlessness. (Above, MA 2016-53-030; below, MG 183-F26-003.)

Northern Pacific locomotive No. 684 had quite the shine and polish after being reconditioned in 1952. No. 684 was originally built by New York Locomotive Works in 1883, and after serving other railroads in the west, it was purchased by the Nez Perce Railroad in 1928 and renamed the No. 4. After a few close calls to joining a scrap heap, No. 684 was retired from service in 1945. Retirement was not quite a glorious adventure, however, and No. 684 sat abandoned in a field for a few years before being sold again 1951. This time, it got the recognition it deserved. (MG 183-F25-028.)

When engine No. 684 was retired, it was not scrapped, but instead sent to Spokane for refurbishment. Once it was finished, it traveled around the West on exhibition. In one location in the early 1950s, people were allowed to enter the cab and get a close look at how trains operate—this aspiring engineer seems to be enjoying the tour. This magnificent historical artifact is still on display in Fargo, North Dakota. (MG 183-F25-032.)

This Northern Pacific passenger car was built in 1883 and retired in 1953, just a couple years before local passenger service began to be discontinued on the Camas Prairie Railroad. The car would have been pulled by wood-fired steam, coal-powered steam, and diesel-powered engines over the course of its life. It was restored to its original condition and displayed by the Northern Pacific until 1974, when it was sold to an amusement company in Michigan. (MG 183-F09-009.)

Union Pacific No. 2037 and Burlington Northern Nos. 2076 and 2077 are making a turn at the wye at Spalding. Managing these kinds of exchanges would have taken a lot of planning, as radios did not become common in trains until the 1970s. Before that time, hilly terrain made the use of two-way radios infeasible. (MG 183-F26-032.)

Maintenance was an ongoing battle in the railroad business. As shown above, a new spur track had to be built to move an engine that had been retired and essentially left in a field. Engineer O'Dell is driving a spike into the rail, and Herb Banks oversaw the work it took to make the engine capable of running. At right, an unnamed employee repairs line damaged from the freezing of the ground. The tool leaning up against the rail is a jack, which was typically used to lift rail cars. Depending on the make and model, jacks can lift from 5 to 15 tons. (Above, MG 183-F25-015; right, MG 457-164-492.)

It is not difficult to see how the Camas Prairie Railroad got its nickname, the Railroad on Stilts. This trestle—the second-longest and tallest on the CPR—is 684 feet long and 141 feet high at its farthest distance from the ground, and is one of the most iconic structures on the entire line. Built from heavy timbers, the gracefully arcing feature is affectionately known as the Half Moon Trestle. It can still be seen from US Highway 95 below it. The 11.5 miles of track along which the Half Moon Trestle is located contain 7 tunnels and 17 trestles and might be the curviest section of the entire route. (MG 183-F15-003.)

The idyllic flatland adjacent to the Snake River served as an occasional home for Indigenous peoples from the Nez Perce and Palouse tribes. This location is called Wawawai after a Nez Perce term meaning "council grounds." White colonizers arrived in 1870 and quickly realized that the fertile ground and warm climate were perfect for growing apples, pears, plums, peaches, grapes, cherries, and berries. (PG 90-10-060e.)

The town of Wawawai, Washington, is on the Snake River west of Lewiston and was a promising agricultural area known for its orchards. The locations marked in this photograph are, 1: the Oregon Railroad & Navigation Company bridge across a creek, along with a road at right under the bridge going to a ferry; 2: a tower supporting a ferry cable; 3: the Wawawai store and post office; 4: the Oregon Railroad & Navigation Company depot; and 5: a fruit-packing building. (PG 90-10-060h.)

The Potlatch Lumber Company was founded in 1903, and a company town named Potlatch was established in 1905. After World War II, the company opened a plywood plant in Lewiston to meet the booming demand for housing. The Potlatch Company remains one of the largest paper manufacturers in the world, and it all began on the Camas Prairie Railroad. (MG 183-F28-003.)

Logging has evolved from mostly involving hand tools and horses or other beasts to maneuver logs through the woods to requiring many men to organize along with the use of rudimentary steam- or gas-powered machines to pick up logs one at a time for ease of transport. Shipping by rail is still the most efficient way to transport large quantities of logs, but modern machines now require only one person to load a whole rail car. (Above, MG 457-165-1024; below, MG 457-163-332.)

Despite being one of the earliest routes, albeit traveled by steamship, and one of the most important for connecting to the larger cities of the West Coast, the Lewiston to Riparia branch became the third subdivision of the Camas Prairie Railroad. These 72 miles of the CPR are also the most desolate, bordered on one side by the Snake River and on the other by the desertlike Channeled Scablands. (MG 183-F20-010.)

At the westernmost point of the Camas Prairie Railroad, this crossing over the Snake River at Riparia was one of the most influential for connecting the Inland Empire to the great shipping ports of the West. From the earliest days of the railroad reaching the inland northwest, Riparia was key in connecting the larger cities of Spokane, Portland, and Seattle. Multiple bridges had to be replaced, and the entire route was moved, to avoid the rising water when the Snake River was dammed between 1962 and 1969. At its peak, the Snake River rose 100 feet above its original level. (Above, MG 183-F20-002; below, PG 99-R-017-005.)

Just because it is nighttime does not mean these trains are resting. Here, Union Pacific Nos. 1236 and 248, along with Burlington Northern No. 1704, are ready to spring into action at the East Lewiston yard. The amount of lumber available to ship and the demand for building material ensured that many lines of the Camas Prairie Railroad ran during all hours of the day and night. One short section of the line that ran only from Lewiston to Orofino was called the Night Logger. It brought empty cars to Orofino each night and picked up cars loaded with plywood and other lumber that were sent to Orofino during the day from the various mills along the route to Headquarters. (MG 183-F19-030.)

The promise of a railroad brought new life to many places in terms of prosperity and population growth. Orofino, which was already a boom-and-bust town thanks to being one of the earliest places gold was found in Idaho, enjoyed quite a bit of new life with the arrival of the Camas Prairie Railroad. Above, the city swimming pool is shown directly adjacent to the tracks. Riparia was another prime example of a town that hoped for tourism. The primarily agricultural town was in a desolate area miles away from larger populations, but the arrival of trains prompted the building of a hotel. (Above, PG 13-7015; below, PG 99-R-017-007.)

The scenic location of the Camas Prairie Railroad has served surprisingly few creative outlets. Here, a fight scene between characters played by Charles Bronson and Archie Moore in the 1975 movie *Breakheart Pass* is being staged on the snowy roof of a train car as it hurtles down the track. All the railroad scenes for *Breakheart Pass* were filmed on the CPR. A station set was built near Arrow Junction, and an accident was staged using a separate track that was constructed to show some cars careening off a cliff near the Half Moon Trestle. Another movie filmed along the route, *Wild Wild West* starring Will Smith, briefly showed a train entering the iconic curved trestle just after leaving a tunnel on the second subdivision of the railroad. (MG 183-F17-001.)

After tracks are abandoned, it does not take long for vegetation to reclaim the land. As early as 1953, older sections of track were already being overgrown. Here, a forester measures regrowth on an old railroad grade near Cow Creek in the Orofino Creek basin, near the fourth subdivision of the Camas Prairie Railroad line that traveled toward Headquarters. (MG 457-160-160.)

Rising costs and farmers seeking alternate shipping routes led to the abandonment of the second subdivision in 2000, with the last train traveling this line in 2002. By 2008, much of the track had been removed, leaving a gash on the landscape. The 2013 Idaho Statewide Rail Plan suggested that the Camas Prairie Railroad's abandoned lines had the potential to be preserved and reused for heritage tourism. (MA 2021-40-001.)

Exploring historic byways is a popular pastime, and the Camas Prairie Railroad is no exception. Roadside markers dot the highway, and frequent pull-offs allow visitors to stop and gaze up into the hills at the railroad's still-visible trestles. Braver adventurers can hike along the route over the trestles and through the tunnels. Unfortunately, the trestles have recently fallen into disrepair, with one burning down in a 2011 wildfire, and with little apparent interest in rebuilding, they can quickly become rather treacherous. (Above, MA 2016-53-005; below, MA 2021-40-002.)

Four

End of the Line

As with many businesses, the Camas Prairie Railroad has had its ups and downs. With a lifespan of nearly a century, the railroad defied the odds surviving economic downturns, conquering multiple terrains with different environmental challenges, and enduring a constantly fluctuating system of governance.

Although the railroad was once the only mode of transportation for both inhabitants of the rural reaches and the products used to support their livelihoods, roads and highways became the preferred method as vehicles became more reliable and affordable, and the improvement of roads meant that more people could drive themselves and not have to rely on the strict schedules and fixed routes of the railroad. The depletion of timber prompted the closure of mills along the route, which also signaled the death of the railroad.

The scenic routes prompted some interest in tourism. Excursion trains on the second subdivision were offered as early as 1960, five years after passenger service stopped on this route. While the first one seemed successful, they did not continue. Another attempt at tourism involved an effort to turn rights-of-way into pedestrian or bicycle paths in a manner similar to the Rails to Trails program that successfully reused many other railroad lines, but this also had its unique challenges. After so many years of neglect, the CPR's trestles and tunnels were in bad shape and would be expensive to make safe. Vegetation and trees have started reclaiming the routes, making it difficult to see where they originally ran. Fires also had an impact on the line, as a 2011 wildfire destroyed one of the large wooden trestles on the Grangeville subdivision. People still attempt to follow the routes on foot or ATV, but they do so at their own risk. Locomotives still have their advantages, however, and trains can still be seen following along the Clearwater and Snake Rivers as they transport products from the Lewiston Mill and grain terminals to larger markets.

HISTORICAL SOUVENIR

Camas Prairie RAILROAD EXCURSION

Lewiston - Grangeville

Two Round Trips

Sunday, May 15, 1960

First Locomotive to arrive in Grangeville from Lewiston, Nov. 8, 1908. First passenger train arrived Dec. 9, 1908.

Excursion Sponsored by

Price: 25c

This six-page brochure from a May 1960 excursion attempted to gin up interest in nostalgia and reliving the historical events of the first train arriving in Grangeville in 1908. Articles speak to the excitement of the railroad inching its way closer, giving updates on its construction and hints about all the celebrations planned for when it arrived. The residents of Grangeville and the rural prairie had great hopes that the railroad would bring opportunities and better connect them with the rest of the world. The brochure also provides interesting facts about events that happened during the years of Camas Prairie Railroad service. (MG 183-F08-001-001.)

As this journey along the Camas Prairie Railroad ends, the authors hope that the visions of steep canyons, dense forests, and desolate desert have been enjoyable and not a discouragement from visiting. Once the agent in charge at the Lewiston depot determines the train is ready to go, it is off to the next destination over the hills, through the forest, and following the river. All aboard! Now, it is time for a well-deserved doughnut break. (Right, PG 91-628; below, MG 183-F21-002.)

Bibliography

Boone, Lalia Phipps. *Idaho Place Names: A Geographical Dictionary*. Moscow, ID: University of Idaho Press, 1988.

Bovey, Byron D., and Ilo-Vollmer Historical Society. *High Line of the Camas Prairie Railroad*. Craigmont, ID: Ilo-Vollmer Historical Society, 2007.

Faris, Charles L., and Don Dopf. *Charles L. Faris: Western Railroad Engineer*. Littleton, CO: Profitable Publishing, 2005.

Hillebrant, Thomas. *Palouse Rails: Granger Railroads of the Inland Northwest*. Charleston, SC: Arcadia Publishing, 2018.

Miller, Garry, and Roz Miller. *The Camas Prairie: Idaho's Panhandle Railroad*. Cheyenne, WY: Union Pacific Historical Society, 2020.

Riegger, Hal. *The Camas Prairie: Idaho's Railroad on Stilts*. Edmonds, WA: Pacific Fast Mail, 1986.

Robertson, Donald B. *Encyclopedia of Western Railroad History*. Caldwell, ID: Caxton Printers, 1986.

About the University of Idaho Library

The University of Idaho Library has grown from a single classroom in the University Administration Building in 1892 to the largest library in the state of Idaho, housing well over a million books and almost 10,000 periodicals in print and online. The library has also served for over a century as an official regional depository of federal government publications, making almost two million government documents available to the public. The library's special collections are an invaluable resource for researchers, providing access to historical photographs, state documents, university historical materials, rare books, digital collections, and the International Jazz Collections—the premier jazz archives of the Pacific Northwest.